CASE STUDIES IN GEOGRAPHIC INFORMATION SYSTEMS FOR INTERNET PORTALS

June 2012

Prepared for:
Office of Planning
Federal Highway Administration
U.S. Department of Transportation

Prepared by:
Organizational Performance Division
John A. Volpe National Transportation Systems
Research and Innovative Technology Administration
U.S. Department of Transportation

Acknowledgments

The U.S. Department of Transportation John A. Volpe National Transportation Systems Center (Volpe Center), in Cambridge, Massachusetts, prepared this report for the Federal Highway Administration's (FHWA) Office of Planning. The project team consisted of Benjamin Cotton of the Transportation Planning Division and Michael Clark of MacroSys. Mark Sarmiento of FHWA's Office of Planning provided project oversight.

The project team would like to thank the Kansas Department of Transportation, the Iowa Department of Transportation, the Kentucky Transportation Cabinet, the New York State Department of Transportation, the Georgia Department of Transportation, the Minnesota Department of Transportation, and the Northeast Ohio Areawide Coordinating Agency for taking the time to talk about their experiences developing GIS portals. The sharing of their experiences and their review of the case studies presented were essential to creating this report.

Contents

1. Introduction 5
2. Case Studies 6
   2.1 Kentucky Transportation Cabinet Interactive Mapping Program 6
   2.2 New York State DOT's Oversize/Overweight Pre-Screening Tool 12
   2.3 Georgia DOT's GeoTRAQS 18
   2.4 Iowa DOT's Snow Plow Portal 22
   2.5 Kansas DOT's KGATE 29
3. Conclusions and Lessons Learned 33
   3.1 Essential Portal Components 33
   3.2 Lessons Learned 34
   Appendix A: Interview Guide 36

Executive Summary

The following report investigates the experiences of transportation agencies in the deployment of Internet-based mapping portals based on geographic information systems (GIS). The report presents background information, a series of case studies, and a summary of conclusions.

As part of this effort, the authors of this report interviewed representatives from five state transportation agencies about their respective GIS portals:

- Kentucky Transportation Cabinet's Interactive Mapping Program;
- New York State DOT's Oversize/Overweight Pre-Screening Tool;
- Georgia DOT's GeoTRAQS;
- Iowa DOT's Snow Plow Portal; and
- Kansas DOT's KGATE.

State DOTs turn to GIS portals in an effort to serve transportation data in a format that is easily accessible through the internet. Although GIS portals limit analytical capabilities typically seen in desktop GIS applications, the web-based portal targets the needs of its users by focusing on identified business processes. Strong data stewardship programs ensure agencies are able to maintain accurate data across multiple disciplines, and as more user groups are accommodated, most GIS teams feel that portals will be an important component of the state DOT spatial data portfolio for years to come.

1. Introduction

For today's transportation agencies and organizations, Geographic Information Systems (GIS) is an everyday tool used in many different capacities. Among its myriad applications, agencies use GIS to manage assets such as roads and bridges, map routes for trucks and buses, track safety incidents, plan for new facilities, and monitor traffic activity. As more transportation agencies collect, store, and manipulate spatial data, it has become increasingly important to ensure that these data are accessible to all feasible users.

In the Internet age, transportation agencies have found new ways to present GIS information to users through online mapping services. Web-based mapping tools have been in existence for over a decade now, and many of the most prominent GIS companies are in the business of creating software intended for the development of web-enabled mapping tools. Many transportation agencies have also made data available for download for use by GIS practitioners outside the agency.

In recent years, technological advancements in web-based mapping and data management are contributing to the deployment of GIS portals by many transportation agencies. Copious amounts of data are able to be stored in-house and queried from anywhere in the world. Access to sensitive data is controllable. More and more professionals are joining the workforce with GIS awareness or extensive GIS experience and programming knowledge. Most importantly, an increasing number of software tools are available to build a dynamic web application. As barriers fall, GIS web portals are more attainable than ever for State departments of transportation (DOTs).

These developments have emerged alongside an increased call for transparency within government. Strategic planning and project-level management are no longer carried out behind closed doors. Rigorous data are essential to making quality decisions in the transportation industry. For this reason, staff members at transportation agencies and the general public must be able to easily access and work with transportation data. Furthermore, budget restrictions facing many transportation organizations nationwide have obligated the development of new tools to carry out these responsibilities with fewer resources.

With the emergence of the GIS portal, transportation agencies are investing in streamlined spatial solutions, fast performance, and intuitive capabilities. GIS portals are quick, easily readable, online geospatial data viewers. The concept is not unlike a vehicle dashboard, allowing viewers to swiftly get the pulse of the situation without much effort.

2. Case Studies

The following case studies on GIS portals complement FHWA's ongoing effort to promote innovative applications of GIS in transportation. In addition to the case studies themselves, this report shares key findings and themes that emerged during the research process, as well as discuss some of the emerging trends and technologies that will likely have an effect on future development of GIS applications for the Internet.

As part of this project, the research team participated in the 2012 Annual GIS-Transportation (GIS-T) Symposium held by the American Association of State Highway Transportation Officials. In addition to learning about current ongoing efforts, the team was able to gather information from the private sector about emerging technologies and software development that will likely have an impact on the web-enabled GIS industry.

The project team also conducted interviews with seven public entities, many with a transportation-specific mission. These agencies were chosen to reflect diversity in the purpose of the portal, specifically in internal versus external use, modes incorporated, and overlying problem that the application was developed to address. Diversity was also sought in geographic distribution, complexity of the application, management of the application (e.g., IT vs. broader staff), user base, administrative support, systems used (e.g., database, GIS, and web application), and data available for analysis. The five agencies profiled in case studies are:

- **Kentucky Transportation Cabinet (KYTC)**—An assortment of individual GIS portals tailored toward specific subject matters. Strong outreach efforts and a reduced mingling of unrelated data within each portal allows for a highly customized and valuable tool for many users throughout KYTC.
- **New York State DOT**—The Oversize/Overweight Vehicle Pre-Screening Tool produces information on height and weight restrictions for roadways, as well as diversions caused by construction activity, to support permitting for oversize and overweight freight movements.
- **Georgia DOT**—GeoTRAQS is a new interactive mapping tool developed by GDOT which enables a wide range of users with one-stop access to much of GDOT's publicly available data. The tool is particularly robust in allowing users to specify their data needs by both subject matter and geographic location.
- **Iowa DOT**—An evolving application with an intricate data collection scheme, the Iowa DOT Snow Plow Portal provides decision-makers with real-time and historical materials usage data to influence maintenance activity during winter weather events. It strives to achieve cost savings by modulating the amount of materials used to treat roadways.
- **Kansas DOT**—KGATE is a collection of 70 feature datasets from a variety of different sources through KDOT and other publicly available sources. It is used primarily to generate custom queries, assist with report writing, and support project-level planning decisions.

Two additional organizations, the Minnesota DOT and the Northeast Ohio Areawide Coordinating Agency, were interviewed for this report but not profiled for case studies. Each interview, however, was instrumental to establishing a context for this report.

2.1   Kentucky Transportation Cabinet Interactive Mapping Program

Background

The Kentucky Transportation Cabinet's (KYTC) GIS Office created its first online interactive mapping tool in 2001. That user-friendly GIS portal created to support the Commonwealth's long-range transportation planning efforts only marked the beginning of KYTC's long-lasting effort to provide the Commonwealth with fast, straightforward, online mapping tools. Over the next decade, KYTC created, hosted, and maintained more than 20 individual mapping portals, all of which were used by a wide range of customers including agency engineers, planners, legislators, private contractors, and the general public.

Today, KYTC's interactive mapping site is a series of unique, interactive GIS portals, each of which acts independently with its own subject matter. The subject is determined based on real-world, task-driven business processes that occur within the Cabinet, and the data within each portal is streamlined to ensure an uncomplicated application that is optimized for the specific tasks of its users. Using ESRI's ArcServer software, KYTC's GIS team opted for a series of portals in an effort to clearly delineate the intent of each tool and avoid cluttering any one portal with excessive data.

Application

The purpose of KYTC's interactive mapping program is to provide as much spatial information as possible by serving that data to its users via interactive online maps. By disseminating this information, KYTC makes data easily accessible with the hope of reducing the number of requests made to the GIS team for transportation data. The audience for the interactive mapping services includes a wide range of users within KYTC, the Commonwealth government, the private sector, and the general public.

At present, the "Maps" page on the KYTC website is a clearinghouse of spatial data that includes numerous interactive mapping portals and printable maps.

Current portals include:

- Active Highway Plan: Displays current Highway Plan
- Bike Routes: Provides official Kentucky bike routes
- Local Roads (Previously County Rural Aid):Offers users information on locally-owned roads
- Environmental Overview: Displays information on environmental documentation and compliance in road maintenance and construction
- Emergency Funding Routes (FEMA\FHWA\ER): Displays routes for FEMA\FHWA\ER funding purposes
- General Highway Map of Kentucky: Displays Kentucky's road network
- Project Plan Archive (1909–Present): Houses an archive of scanned KYTC project plans: 1909–present
- Roadway Photo Viewer: Includes a photo log of the Kentucky road system
- Right-of-Way Monument Map: Provides information on KYTC's right-of-way monuments
- Rural and Secondary Roads: Displays rural and secondary road aid allocation
- KYTC Traffic Counts: Indicates traffic count locations, AADTs, functionally classified roads
- KY CORS and Geodetic Control: Displays current KY CORS stations, CORS Baselines, and other Geodetic Control

Between 2001 and 2011, KYTC's GIS web tools were built using ESRI's ArcIMS software. In 2011, after the program had grown to more than 20 mapping sites, the agency began the migration from ArcIMS to ESRI's ArcGIS Server. Given its history working with ESRI's GIS products, ArcGIS Server was a natural progression for KYTC as the GIS team sought to improve the performance of its web mapping services.

Data Stewardship

Rather than combine all spatial transportation data into one GIS portal, the KYTC GIS team opted to present its data in a series of portals. The primary benefit of this model is that each data portal is self-contained and straightforward to use and can be tailored to the customers' specific needs. Site owners are engaged in a collaborative dialogue to identify how the mapping tool with address their business needs, providing a focused solution that minimizes the decisions that a customer has to make by streamlining the data and focus topic of the various sites.

A potential drawback with this model is that the portals do not allow on-the-fly addition of data. Based on its understanding of how people use the portals, however, the KYTC GIS team believes this is one of the portals' advantages. According to the team, users are rarely looking for complex spatial analysis. Rather, these maps are being used as part of business processes that require rapid performance and clear data interpretation. The team believes that with too many "bells and whistles," the utility of the tool may be lost on the most important user groups. Furthermore, downloadable data is available from KYTC on request for GIS users looking to perform more robust analyses.

Figures 1, 2, and 3 are a series of screen captures of the KYTC Highway Plan portal. The typical user's data procurement process is likely to include opening the portal and viewing the statewide map, zooming in on a particular location, and then identifying a feature for which more information is desired.

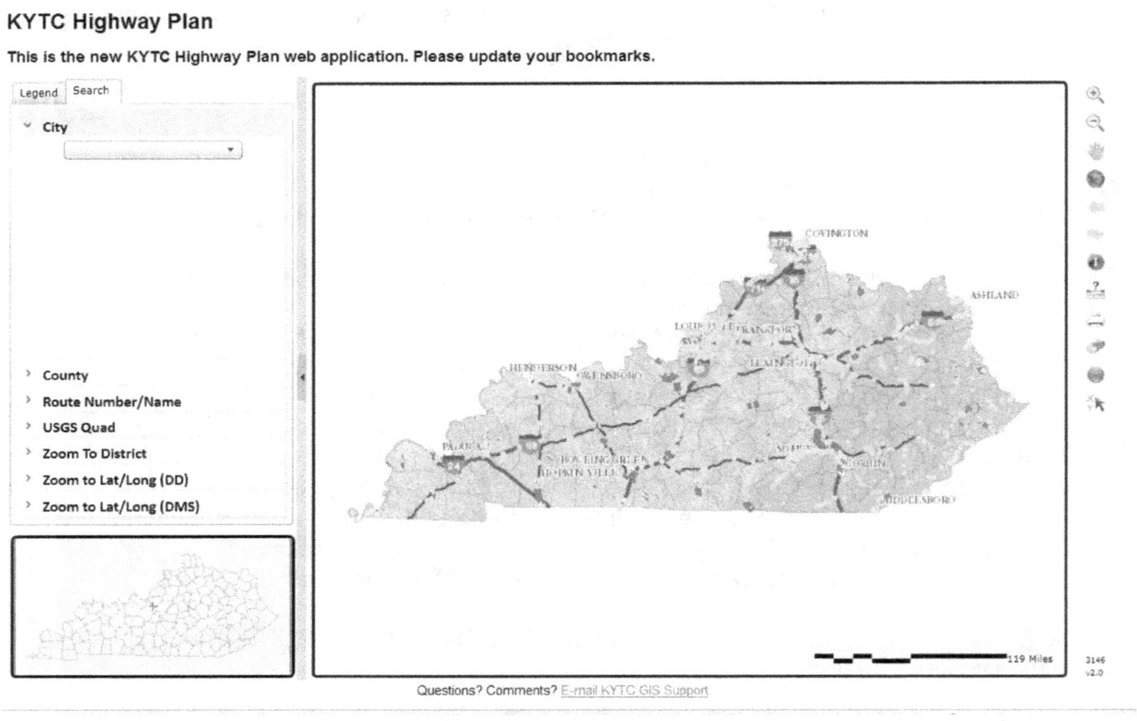

Figure 1: KYTC Highway Plan portal
A statewide view of the KYTC Highway Plan, completely zoomed out. Green lines represent planned projects, while red lines indicate authorized or awarded construction projects.
Source: http://maps.kytc.ky.gov/SYP/

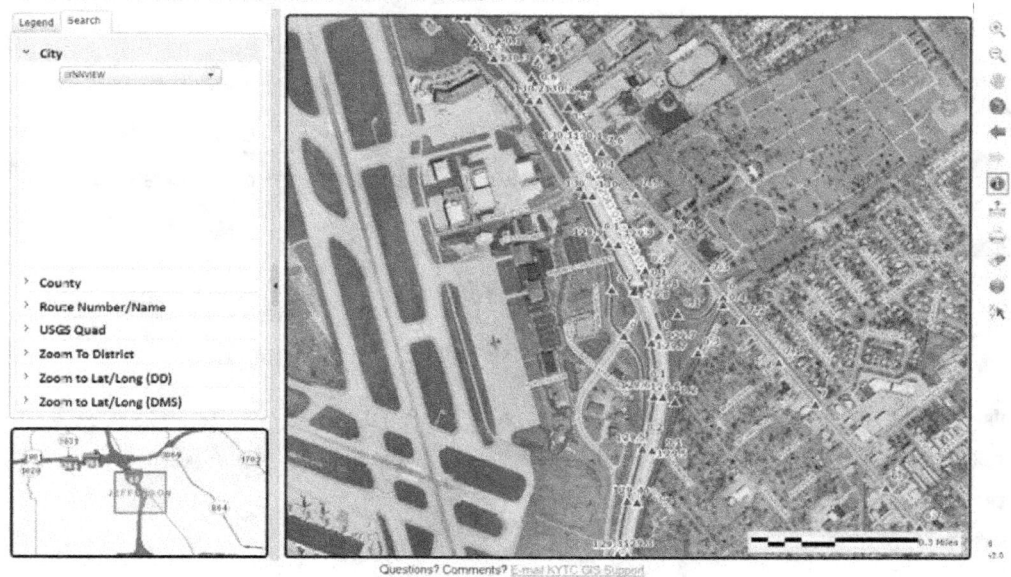

Figure 2: KYTC State Highway Plan
A localized view of the KYTC Highway Plan. Green lines represent planned projects, and orange lines indicate bridges.
Source: http://maps.kytc.ky.gov/SYP/

Figure 3: Example of preconstruction documentation

Documentation for each transportation project is spatially enabled and attached to each project. Document includes a project description, financial information, and important dates.

Source: http://maps.kytc.ky.gov/SYP/

Each individual interactive mapping portal that KYTC hosts has a designated "owner" who is responsible for data maintenance. In some cases, such as for road centerline data, the KYTC Office of Planning maintains the data, automatically extracting the data from its production database for inclusion in the portals. In other cases, site owners are responsible for monitoring data development and notifying the GIS team when the portal needs to be refreshed. Most importantly, the GIS office emphasizes that it is not in the business of creating data; rather, it facilitates and coordinates the viewing of data as needed by its customers. As a result, once a GIS portal is configured and launched, very little data maintenance falls into the hands of the KYTC GIS team.

Site owners are also responsible for deciding what data to include in each portal. Sometimes the data that they need is already on hand; other times, as is the case with orthoimagery, sources will be tapped within the Commonwealth's Division of Geographic Information, part of the Commonwealth Office of Technology, with whom the KYTC GIS team has a longstanding working relationship.

Data Stewardship

The KYTC GIS Team is part of the Cabinet's Office of Information Technology (IT), but given its specialized role and growing influence on the transportation industry, the GIS team is a branch office within IT. With a staff of six permanent employees, each with his or her specialty, the GIS team develops a set of goals each year that it hopes to accomplish. These goals are directly tied to the Office of Information Technology's goal, which are in turn tied to those of the agency as a whole. Each year, the goals are evaluated both at the team and executive levels to assess performance and adjustments that need to be made in subsequent years.

While the current GIS team is relatively small, expectations that the role of spatial technology will continue to grow within KYTC may result in an expansion of the GIS team in the near future. Recent institutional changes show that fundamentals are shifting to account for increasing needs in spatial technology, and all indications are that as demand for spatial data increases, KYTC will continue to support the work that the GIS team is accomplishing. For example, GIS experience is now a job requirement for the majority of KYTC's incoming engineers, representing a fundamental change in mindset as the agency looks to the future.

Outreach and Training

Promoting the capabilities of its transportation portals is an important priority of KYTC's GIS team. It offers training opportunities throughout the state as frequently as once per month, with a current goal of visiting every district in the state over the next 18 months. In many cases, training focuses on basic tasks related to using the GIS portals to view and make maps. The team also offers more advanced training opportunities for geospatial analysis. The idea is that the higher the ability of its users, the more opportunity to develop innovative tools for them to use.

The GIS team believes face-to-face outreach efforts are very important, particularly as they relate to users in the Cabinet's district offices. On-site visits reinforce the mission of the team, as well as dispel any tension that may exist between the central office and the district offices.

Performance metrics

The success achieved by KYTC's interactive mapping program is not measured through web statistics, rather, the GIS team uses feedback from surveys to develop and adjust its annual goals and five year plan. The team takes advantage of every opportunity to collect information via survey from its customers: an annual survey is provided to its regular customers; feedback is solicited at the end of all training courses; project-specific surveys are given to workers and stakeholders; and a survey is distributed at an annual statewide GIS user conference that the GIS team holds.

According to the team, survey feedback is crucial to the decisions that are made by the GIS team, and the anonymity often sheds light on issues that may otherwise not be brought to the team's attention. For example, the GIS team is currently working to deploy handheld electronic devices for field data collection. When users were visited on-site to discuss any problems with the devices, no major issues arose. However, a post-project survey indicated that many users were dissatisfied with the screen size and glare. Based on the results, the team is now investigating alternative mobile devices with larger screens with anti-glare capabilities.

While the team does receive suggestions for improvement through the survey process, the vast majority of survey responses indicate that GIS initiatives within the Cabinet are beneficial to the users and their work processes.

Challenges and Lessons Learned

In general, the KYTC GIS team has been able to accomplish its goals without facing too many major obstacles. Like most technology-based efforts, however, simply keeping up with industry innovation is an ongoing challenge. Software companies are continually upgrading products and adding new capabilities. Keeping track of these developments is never ending and can be a minor source of frustration. According to the team, there are times when it seems that as soon as a KYTC tool is deployed, the software company releases a new version of the software upon which the tool is built.

Software upgrades may present a good opportunity for making changes in design and content, but because the GIS team cannot know each and every user of their tools, they cannot always be sure that changes, updates, and/or improvements to the tool are serving the needs of everyone in the user group. Some areas of the Cabinet had even developed work processes based on the interactive mapping portals, of which the GIS team was not aware. When the interactive mapping program migrated from ArcIMS to ArcServer, some of these business processes were unexpectedly interrupted, requiring the GIS team to scramble to get the service up and running again.

The final challenge facing the GIS team is defining the role of the GIS team within the KYTC Office of IT, as well as their relationship with the Division of Geographic Information in the Commonwealth's Office of Technology. While managing server access, space, and maintenance will always be a challenge, the GIS team is generally happy with the way GIS is becoming integrated into KYTC's IT programs.

Future Plans

Generally, KYTC's GIS team would like to add more tools and analytical capabilities to its GIS portals. At the same time, the GIS team is cautious about overwhelming its users. Given the ways the team understands the tools and how they are being used, any improvement will be thoroughly vetted to ensure that it fits into the task-driven nature of the program.

With that said, several new GIS portals are in the works for KYTC. One tool, currently in its final stages of testing, is a permitting portal. Internal to KYTC, the permitting portal will reside in a SharePoint environment where level of access to data is easily managed. The tool will capture those projects that are on the verge of production and display them on a basemap. Then, when a user clicks on a project, all acquired permits will be available for viewing. The permit portal will be seamlessly integrated with wireless, handheld field data collection devices.

In developing its interactive mapping program and series of GIS portals, KYTC has created a place on the web that solves the transportation data needs for a wide range of customers. With a focus on coordination and facilitation, the GIS team offers an array of clear, user-friendly mapping tools that are relatively easy to build and maintain. As its audience continues to grow, the KYTC GIS team plans to continue to progress toward the ultimate goal of imbedding spatial information into every aspect of KYTC's work.

2.2   New York State DOT's Oversize/Overweight Pre-Screening Tool

Background

The New York State Department of Transportation's (NYSDOT) Oversize/Overweight Vehicle Pre-Screening Tool allows those seeking oversize or overweight vehicle permits and NYSDOT's permitting office to check on up-to-date bridge and highway weight restrictions for oversize and overweight freight movements. First used in 2006, the tool has enabled applicants and permitting officials alike to refer to this information in a timely manner.

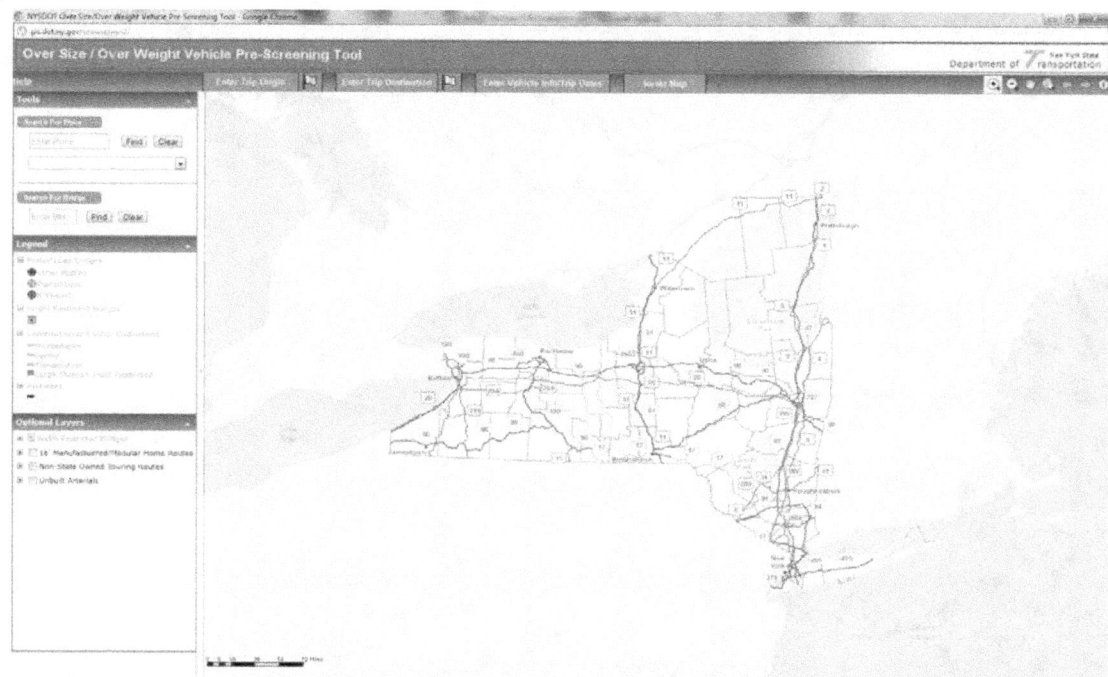

Figure 4: NYSDOT Oversize/Overweight Tool

The full-zoom home screen of the NYSDOT Oversize/Overweight Vehicle Pre-Screening Tool shows the major highway corridors for all of New York. Legend and layers are available on the left side of the screen.

Source: http://gis.dot.ny.gov/osowscreen2/

Application

The Oversize/Overweight Pre-Screening Tool was developed with two purposes in mind: to provide transportation users of state roadways with up-to-date legal vehicle routing, and to serve as a resource for the permitting office to process applications. For example, the tool offers users access to permitted vehicle clearance information for height-restricted bridges. Other selected bridges feature posted load limits. Users can also find the locations of R-posted bridges, which do not allow travel of vehicles over posted weight limits. Bridge closures are also listed. Each of these characteristics is displayed via different color coding.

Construction or maintenance projects on roadways and bridges are also provided. Users can find information on the travel direction of restrictions, as well as vehicle height, width, length, and weight restrictions during the restriction event, the dates in which the restriction will be in effect, and comments and special requirements for transportation users. Those projects that restrict use of large trucks are color-coded on the mapping application, along with the locations of truck-restricted parkways. Users also have the option of displaying selected layers relating to width-restricted bridges and pre-defined manufactured/modular home corridor routes.

Using this information, application users enter vehicle characteristics relating to the width, height, length, and overall gross weight. They also can enter their travel dates. With this information, the Pre-Screening Tool shows only those bridges and roadways for which a restriction is active. Although the tool does not currently identify restriction sites along a pre-defined route, users can develop the most efficient route for their trip with the data provided. Users are encouraged to visit the site frequently due to the changing nature of some of these restrictions

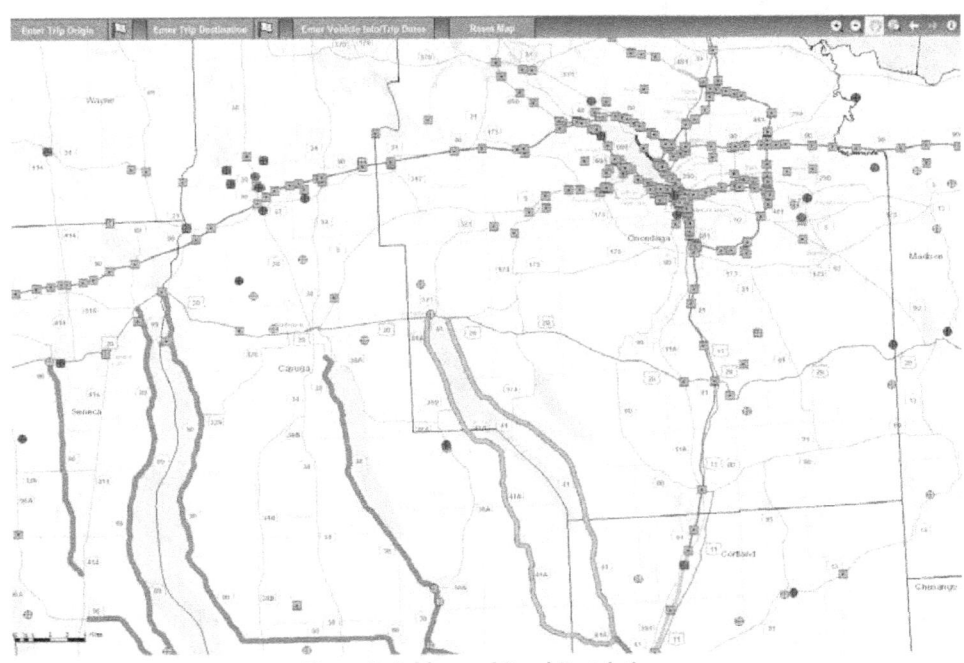

Figure 5: Bridge and Road Restrictions
The Pre-Screening Tool shows bridge and roadway restrictions in the area southwest of Syracuse, NY. Orange roadways signify construction or maintenance activity and red roadways are restricted for large trucks.
Source: http://gis.dot.ny.gov/osowscreen2/

Prior to the tool's development, permit seekers did not have a way to refer to bridge and roadway restrictions on state roadways to select an appropriate route. Thus, NYSDOT rejected many permit applications. With the Pre-Screening Tool, permit seekers can see the location of bridge and roadway restrictions throughout the state and formulate a route to avoid these restrictions. This has resulted in more permits being granted on the first application attempt, lessening the resources required from the transportation industry and NYSDOT alike.

Additionally, the permitting office's process for keeping track of restriction data on roads and bridges before the Pre-Screening Tool was antiquated and labor-intensive. A large road map on the wall of the permitting office displayed the location of restricted roadways and bridges. Ensuring the accuracy and completeness of these data was a time-consuming process. Moreover, additional resources were required to review failed applications that came in part from not providing this information to customers.

By feeding the latest restriction data into the Pre-Screening Tool, NYSDOT staff can more adeptly refer to this information in processing permit applications. Automated nightly updating of roadway and bridge activity has eliminated the need for staff members to manually report these events. Furthermore, NYSDOT has adapted the Pre-Screening Tool over time to respond to the needs of the permitting office. For example, a consultant involved with the development of the Pre-Screening Tool has worked with NYSDOT to improve processing speeds and add a geocoding capability to the application. These improvements have helped the Pre-Screening Tool realize comprehensive usage within the permitting office, transforming the way in which the office conducts business.

Technical Specifications

The Pre-Screening Tool was first introduced in 2006 using ESRI's ArcIMS. Working with a local consultant, NYSDOT was able to craft an application that could be operated and maintained within desktop tools already in use at NYSDOT. It is currently operated through the .NET

Framework. Since implementation, little has changed about system capabilities or technical specifications other than a transition to ArcGIS Server 9.3.1 in 2009. The agency is currently exploring migration to ArcGIS Server 10 for all NYSDOT GIS operations.

Data Stewardship

At the heart of the Pre-Screening Tool is the highway and bridge restriction data that NYSDOT engineers supply. These data, which can change frequently, feed into the Pre-Screening Tool through larger NYSDOT data management processes. The processes for linking information from these databases to the Pre-Screening Tool are defined below.

NYSDOT staff members have used bridge height and weight restriction data since before the Pre-Screening Tool. Collected and compiled every two years by engineers, the height and weight data are stored in an Oracle database and linked to another GIS database of bridge locations. A separate GIS database, which pulls the restriction information from the Oracle database, features bridge location data to be used in the portal. These characteristics are collectively displayed in the portal.

Engineers can also enter highway restriction data, such as that related to maintenance and construction events, into a separate ArcGIS desktop program. Unlike the enterprise data management system for bridges, this tool was developed specifically to allow this data to be coded into the Pre-Screening Tool. Proper use of this system requires engineers in regional offices to code the latest restriction data, which are collected on an as-needed basis and compiled into GIS. The Pre-Screening Tool development team then pushes this information to an Oracle database for eventual integration with the portal software.

A customized script uploads the latest data from the two Oracle databases on a nightly basis, ensuring that the information contained within the Pre-Screening Tool is up to date. The automated nature of the process minimizes the need for oversight of the system, freeing up staff time for other purposes.

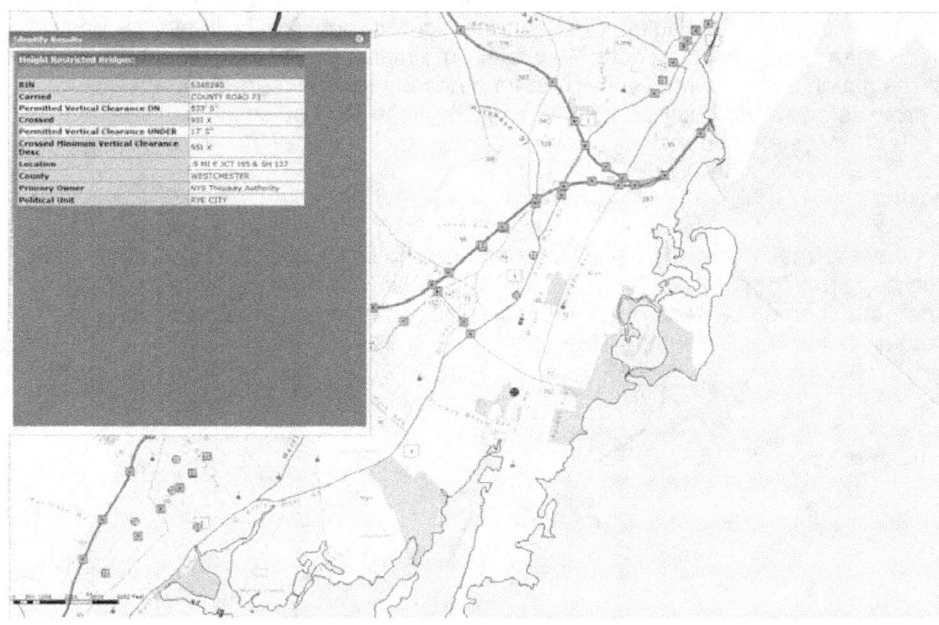

Figure 6: Height restriction data
Height restriction data is shown for a bridge in Westchester County, NY. In addition to the maps, a table provides details about permitted vehicle dimensions.
Source: http://gis.dot.ny.gov/osowscreen2/

Outreach

Ensuring the Pre-Screening Tool continues to meet the needs of the transportation industry is a critical consideration for NYSDOT's permitting office. Although the Pre-Screening Tool was designed with the use of the permitting office in mind, many of the efficiencies it has helped achieve have been made possible through permit applicants using the tool to reduce the likelihood of a rejected permit application.

Intermediaries often submit applications on behalf of permit seekers. Before the tool was first released, many of these intermediaries and others regularly seeking oversize/overweight vehicle permits from NYSDOT were given a chance to preview the tool and provide feedback for improvement. This outreach provided NYSDOT with the opportunity to educate the industry on how to best use the tool, as well as publicize the availability of the tool to assist with permit applications. Although use of the Pre-Screening Tool is not required for applicants, NYSDOT believes a large number of applicants recurrently use the tool to identify bridge and roadway restrictions. This is based on the reduced number of rejected permit applications.

Benefits and Challenges

The primary benefit of the Pre-Screening Tool is the shortened process for oversize/overweight vehicle permitting. Before the launch of the application, applicants had no way to reference the location and nature of potential restrictions along their desired route of travel without contacting NYSDOT and local municipalities directly. Without this information, applicants were more likely to request permits for routes that could not be travelled. Subsequent requests were subject to the same knowledge limitations. This would increase the time that both applicants and NYSDOT staff members would spend creating and reviewing applications. Making restriction data readily available has reduced this burden significantly for each party.

Within NYSDOT, the Pre-Screening Tool has replaced a labor-intensive process of collecting and inventorying current restriction data. Aside from the initial investment to launch the application, which involved working with a local consultant and conducting outreach with its customer base, NYSDOT is relatively "hands-off" when it comes to system maintenance. The nightly script that uploads the latest restriction data minimizes the staff time necessary to ensure data are accurate. The application's processing speed has been suitable for the permitting office since it was upgraded shortly after the system's launch.

Performance Metrics

NYSDOT does not have formal measures by which to assess the tool's effectiveness. Anecdotal evidence, however, suggests that the tool has saved NYSDOT and the transportation agency time and labor. The permitting office sees more successful applications on the first attempt than prior to the introduction of the Pre-Screening Tool.

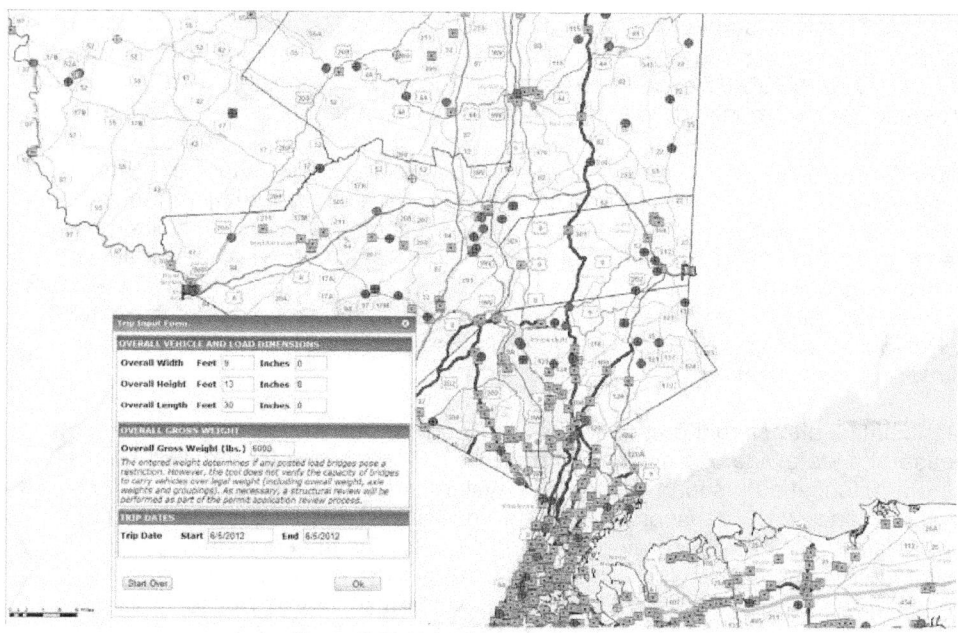

Figure 7: Vehicle dimension entry form
Users can input entry and exit points, as well as the dimensions of the vehicle. The application shows
potential restrictions for a vehicle 9 feet wide, 13 feet high, and 30 feet long.
Source: http://gis.dot.ny.gov/osowscreen2/

Lessons Learned

The success of the Pre-Screening Tool has depended from the outset on good communication among staff members at NYSDOT and between NYSDOT and the transportation industry. Although use of the tool has been compulsory for permitting staff members, there was a learning curve and adjustment period for these users. By listening to feedback, NYSDOT and its consultant were able to update the application to better fit the needs of the permitting office. This included improving processing speeds and adding a geocoding capability to allow users to zoom in to route origins and destinations. Today, the Pre-Screening Tool is highly regarded by the permitting office.

Future Direction

While NYSDOT feels largely satisfied with the positive effect that the Pre-Screening Tool has had on the transportation industry and internal operations, it is also strongly considering the future direction with which to take the tool. At issue is the consideration of whether new capabilities can be introduced to the tool to further maximize internal and external efficiencies. These issues must be viewed in the context of the agency's pending migration to ArcGIS 10 and integration of the tool with other agency projects.

NYSDOT is interested in evolving the tool to become a "one stop shop" for oversize/overweight vehicle permitting. To achieve this goal, permit seekers would need to be able to have their proposed route of travel verified against system data and complete and submit an application form online. Achieving this goal would require that the Pre-Screening Tool become more dynamic in the features it offers to permit applicants. Chief among these new capabilities is a routing tool for users to interactively find the most efficient route for their travel while avoiding roadway and bridge restrictions.

The consideration of a more comprehensive permit automation system drives the future development of the Pre-Screening Tool. Such a system would reduce the time needed for applicants and NYSDOT staff to identify and verify acceptable routes of travel. Staff members are currently focused on the designing and developing the system, which will feature an interactive restriction identification tool. NYSDOT expects to issue a Request for Proposals soon, but until then the Pre-Screening Tool will continue to serve the needs of its internal and external users.

With ArcGIS Server 10 available, the discussion of the Pre-Screening Tool's future direction must also address the direction of GIS usage throughout NYSDOT. Since the tool's implementation in 2006, NYSDOT has reacted positively to its external use and recognized the efficiencies it has brought to the permitting office. The idea use of GIS as a resource to improve other aspects of NYSDOT operations is gaining momentum. The consideration of how NYSDOT will utilize ArcGIS 10 has led to much critical thinking from the agency on how to best implement and deploy the technology. The Pre-Screening Tool development team is also analyzing which API would be best for the application's next iteration.

Another NYSDOT initiative attracting the attention of the Pre-Screening Tool development team is the agency's release of a statewide 511 system. The restriction data at the forefront of the Pre-Screening Tool are a potential data source for the 511 system since road and bridge closures and diversions are of significance to all transportation system users. With this information constantly changing, the methods by which it is integrated into the 511 system can be streamlined to achieve cost and labor savings. Additionally, opportunities may exist for NYSDOT to share their data with third parties through the 511 system, such as navigation software providers. Opening up this data for more general usage holds the potential to reduce traffic incidents and improve travel conditions by notifying motorists earlier of maintenance and construction activity.

2.3   Georgia DOT's GeoTRAQS

Background

The Georgia Department of Transportation (GDOT) has employed internet-based mapping tools since 2001 to help disseminate transportation information to users within GDOT as well as the general public. Its first tool, the Transportation Explorer (TREX), was replaced in 2012 by the Geographic Transportation Reporting Analysis and Query System (GeoTRAQS), an enterprise GIS portal designed to disseminate key GDOT data to a wide range of users. GeoTRAQS, which was developed in nine months, incorporates SharePoint-managed enterprise data, mapping dashboards, and web services into an intuitive user interface driven by geographic extent.

Application

With GeoTRAQS, Georgia DOT aims to improve the look and feel of its online mapping tool, incorporate scalability without compromising performance, and produce data-driven results for its users. Users within GDOT make up a large component of the GeoTRAQS audience, but the agency strives for transparency and aims to make as much of GDOT's spatial data available to the public as possible. The map's straightforward design is familiar to users of mainstream online mapping applications, and the intuitive interface includes a graphic toolbar and on-the-fly panning and zooming.

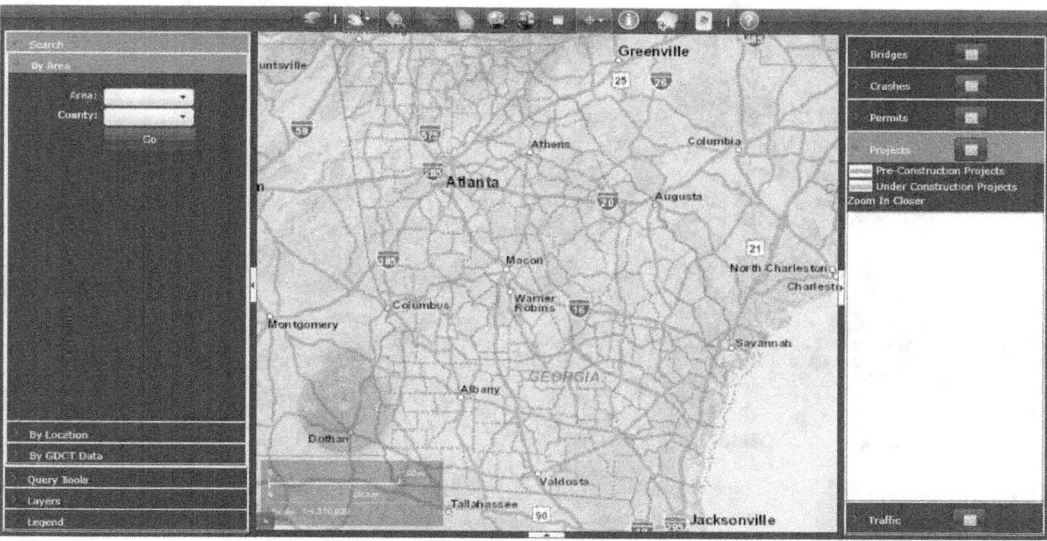

Figure 8: GeoTRAQS home screen
The full-zoom home screen of GeoTRAQS shows major highway corridors in the state. The left and right side of the screen are filters that allow for specific queries.
Source: www.dot.state.ga.us/maps/geotraqs/Pages/default.aspx

More than fifty different layers are available for users to toggle on and off within GeoTRAQS. Key datasets include:

- **Traffic Information** – AADT traffic count locations and reports;
- **Crash data** – location, vehicle information, injuries/fatalities, and accident reports;
- **Permits** – location, documentation;
- **Construction projects** – basic project information, status, description, financial information; and
- **Bridges** – weight restrictions, photographs, and Bridge Inventory Data Listings.

With GeoTRAQS, GDOT equips users with all available data and enables the creation of maps and reports based on their needs. The tool has an advanced search system that allows data to be filtered through an accordion-style format or queried using more advanced techniques such as spatial analysis (buffers, proximity) or feature attribute data. This concept allows a user to use multiple techniques that arrive at the same results; one can freely zoom and pan a map that will allow clicking to identify items of interest, or one can use the system of dropdown menus and search fields to isolate areas via words and text.

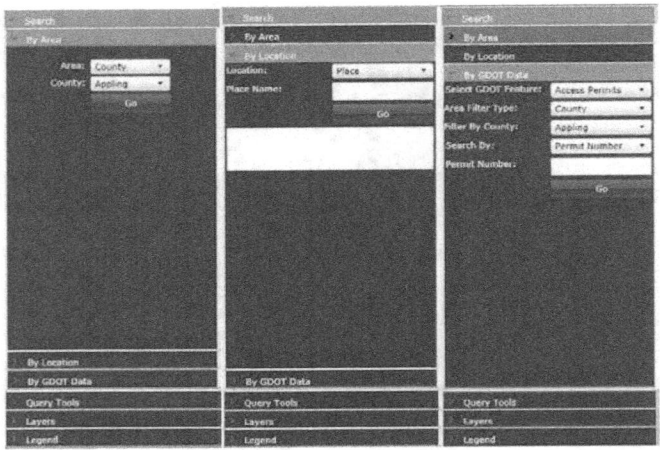

Figure 9: GeoTRAQS drop down menus
Accordion-style drop down filters allow users to isolate areas of
interest by geography and topic.
Source: www.gis-t.org/files/FeU6x.pdf

GIS dashboards also play a role in GeoTRAQS, specifically for bridge data and construction projects. The dashboard is designed to provide all important data about a bridge or a construction project on a single screen. The information can be processed in a glance, quickly printed, or exported to carry into the field on a handheld device.

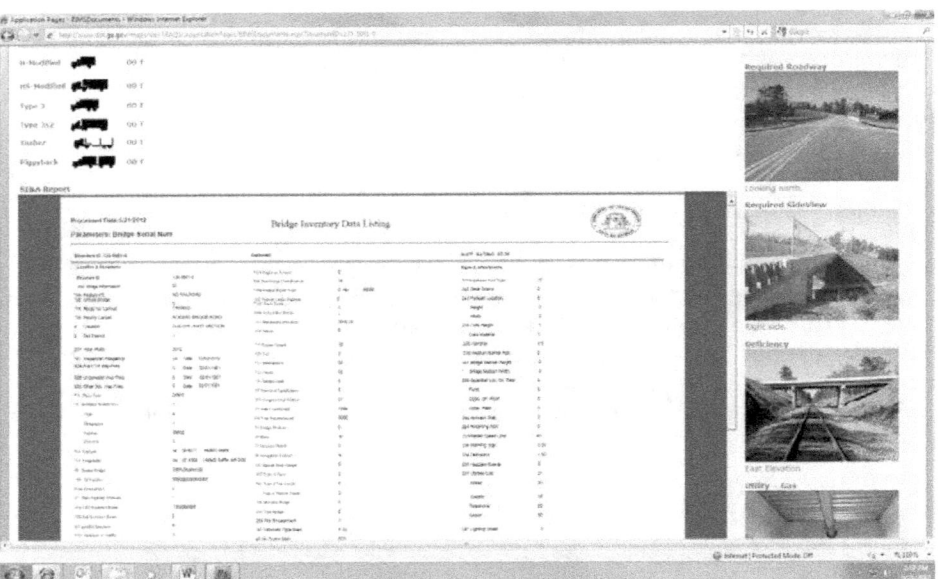

Figure 10: Example of Bridge Dashboard
A bridge dashboard compiles all major bridge data for display on one screen. Elements include the
Bridge Inventory Data Listing, photographs, and vehicle size/weight restrictions.
Source: www.dot.state.ga.us/maps/geotraqs/Pages/default.aspx

Fundamental to GeoTRAQS is an enterprise architecture ensuring that spatial transportation data is able to be managed regularly and efficiently. The architecture consists of four principle components that are the foundation of GeoTRAQS:

- The content management system is housed in Microsoft SharePoint 2010, which the GIS team finds to be a good fit for its particular needs. Data can be uploaded and managed without requiring the skills of a database administrator
- The mapping functionality runs on ESRI's ArcGIS Server 9.3.1 with Microsoft's Silverlight .NET API.
- Reporting and business intelligence is accomplished with SAP Business Objects. This allows reports and charts to be linked to and/or embedded within the tool.
- Several GDOT applications are linked from GeoTRAQS, including TransPI, a project information query tool, and GA STARS, the annual statewide traffic counting program.

Data Stewardship

In most cases, if GDOT data are publicly available, they are able to be displayed in GeoTRAQS. At a database level, however, there is sensitive information that cannot be displayed. SharePoint allows multiple levels of access, and GDOT has organized an executive data governance committee that is responsible for making decisions about data inclusion and potential encryption needs.

GDOT also emphasizes the importance of data stewardship and the concept that the GIS team exists to publish information, rather than create it. As such, each data layer that the GDOT GIS team publishes must be supported by a subject matter expert who can answer technical questions from users, create (or facilitate the creation of) new data when necessary, and monitor any additional maintenance requirements. The goal for data stewardship is to seamlessly integrate data maintenance with data dissemination so the end user does not suffer downtime or inaccurate data.

Support

The current workforce at GDOT includes team leaders who are direct GDOT employees, with a working staff comprised mostly of on-site contractors. Moving forward, GDOT aims to increase the size of its contracted workforce. For the GIS office, this is generally viewed as a positive trend that allows the office to be flexible as workforce needs adjust.

While contractors are able to satisfy the technical needs of GeoTRAQS, there is a need for more staff dedicated to customer support. These employees would focus on providing support to end users, informing them of GeoTRAQS features and how it can be effectively incorporated into business processes. Additional staff members are also in demand to provide technical support to GDOT workers gathering data in the field, as they are the group that enters the data into the system.

Outreach

GDOT finds that questions about GeoTRAQS follow similar themes and often can be answered with content that is already prepared. To accomplish this, they make training materials available to users on the SharePoint site. Training content may come in the form of written instructions, but the agency has also found that video tutorials are helpful when on-screen processes need to be shown. By creating these videos, GDOT is able to eliminate a significant amount of time and cost dedicated to scheduling training and related travel. The agency has also established an ESRI Virtual Campus subscription that provides GIS training classes through the web on an as-needed basis.

GDOT recognizes a need for human interaction when seeking successful training results, and the ESRI Training Pass program helps to fill this need. The year-long program gives GDOT access to certified ESRI instructors who will travel to sites throughout Georgia to facilitate online classes. The program eliminates the need for users in various regions to travel to Atlanta to receive training.

Performance Metrics

GeoTRAQS is a new application, and currently GDOT does not have any formal plan for measuring performance. However, Google Analytics and server side monitoring are being used to provide some indicators. For similar past efforts, GDOT has developed and distributed surveys to internal user groups to learn about successes and improvement needs. Currently, GeoTRAQS has an email distribution list of approximately 400 users, an online discussion forum, and a SharePoint team site. According to GDOT, not all GeoTRAQS users are entirely comfortable yet with social media as a means for communicating or contributing to the agency GIS knowledgebase.

Future Direction

Having been deployed in early 2012, GDOT has several visions for GeoTRAQS in the coming months and years. Generally, the agency wants to improve basemap cartography and performance with a cached tiling scheme, separate public and private services, convert to service data object geometry, and migrate to Web Auxiliary Sphere WGS84. GDOT also wants to create a more personalized experience for its users, by which a user's log-in credentials allows GeoTRAQS to recognize the user, direct the user to the data of interest, and save user preferences for a return visit. Finally, the agency would like to streamline the portal by combining working layers with the map legend.

In the future, GDOT would also like to make GeoTRAQS compatible with TerraGO GeoPDF, a solution developed by Georgia-based TerraGo Technologies that allows spatial data to be packaged in Portable Document Format (PDF). By developing this capability, many processes could be performed on GDOT's data by users without traditional GIS software, including viewing coordinates, changing coordinate systems, measuring lengths and areas, connecting to GPS, and exporting to ESRI Shapefiles.

Finally, GDOT hopes to use GeoTRAQS in conjunction with ESRI's GeoPortal, an open source product that facilitates greater sharing and interoperability of web-enabled spatial data. The GeoPortal works with a data catalog that is compliant with the Open Geospatial Consortium, an international standards body working to empower technology developers to make spatial information accessible with all kinds of applications. The concept aligns with GDOT's vision for the future that transportation data from Georgia will be part of a global network of geospatial information that is universally accessible and interoperable.

2.4 Iowa DOT's Snow Plow Portal

Background

The Iowa Department of Transportation (DOT) is currently developing a GIS-based Snow Plow Portal to assist the agency in asset management for winter operations. Utilizing trucks equipped with data-collecting sensors, the agency can track information pertaining to road conditions and material usage to assess real-time operations during a winter weather event. Although the tool is still in its infancy in terms of influencing resource management within the agency, the robust processes for data collection, data storage, and data analysis using customized web portals show the potential for the tool to transform the agency's handling of winter operations.

Application

Owing to its location in the Midwest, Iowa can be subject to many weather events each winter, including heavy snowfall and sub-freezing temperatures. On average, 40 inches of snow fell each winter between 2006 and 2010. These conditions necessitate Iowa DOT to maintain a fleet of over 900 vehicles to treat roads for snow clearance or skid resistance. Over the 2006–2010 timeframe, Iowa DOT trucks dumped on average over 35,000 tons of sand and 230,000 tons of rock salt on nearly 24,000 state road lane miles each year.

Winter weather operations have required a budget of approximately $40 million annually over that time. Historically, the DOT only had a modest awareness of how well this budget was being used. It was not well-understood how its fleet was being deployed during a weather event and whether materials were being used efficiently. The process for record-keeping was laborious and difficult to evaluate. With the agency's overall budget becoming increasingly strained in recent years, compounded by recent increases in the prices of treatment materials, efforts to achieve cost savings needed to be considered.

The Snow Plow Portal aims to allow the agency to visualize and manage fleet movement and material usage by vehicles in operation during a winter weather event. The tool will allow supervisors to direct the fleet and make decisions using the latest information, as well as replace the paperwork-intensive process of reporting materials use.

At the heart of the Snow Plow Portal is a rigorous procedure for collecting, storing, and displaying real-time data for decision-makers. Iowa DOT trucks have been equipped with external hardware and on-board radio modems to measure and report material usage and road and air conditions. Iowa DOT can pull this information into its databases and display it in a custom-made portal for decision-makers. Although technical issues have prevented the system from being rolled out completely, the quality and speed at which data can be made available for supervisors and other DOT officials is anticipated to present strong opportunities to improve the management of agency fleet. The flexibility of the web portal component makes these data accessible to users of all skill sets over multiple platforms.

Technical Specifications

There are two central technical components to the Snow Plow Portal—the real-time data collection and the processing of these data for agency use.

Iowa DOT selected Location Technologies Incorporated (LTI) as the vendor responsible for the collection of field data and synchronization of that data to the correct location and time that they were observed. To carry this out, each vehicle is equipped with sensor devices to collect data related to conditions, events, and materials application. These data are transmitted through an in-vehicle radio modem to LTI's internal database.

Connecting directly to the LTI database, Iowa DOT pulls this information into its Oracle Spatial database. This occurs on both a real-time (every minute) and historical (every 30 minutes) basis whenever a vehicle is in the field. Real-time truck location information is collected on a continuous basis throughout the course of a weather event. Over time, a "crumb trail" can be created showing the presence and intensity of maintenance activity on roadways.

Due to the enormous quantity of data collected, material usage and condition information must be processed more intermittently. Iowa DOT uses algorithms to pull and aggregate data into exploitable segments. The original goal of the agency was to ping vehicles every 15 minutes to see the status of materials being dumped, but this created a data overload that neither LTI nor the agency had the bandwidth to accommodate, especially in the event of storms requiring utilization of a majority of the agency's fleet.

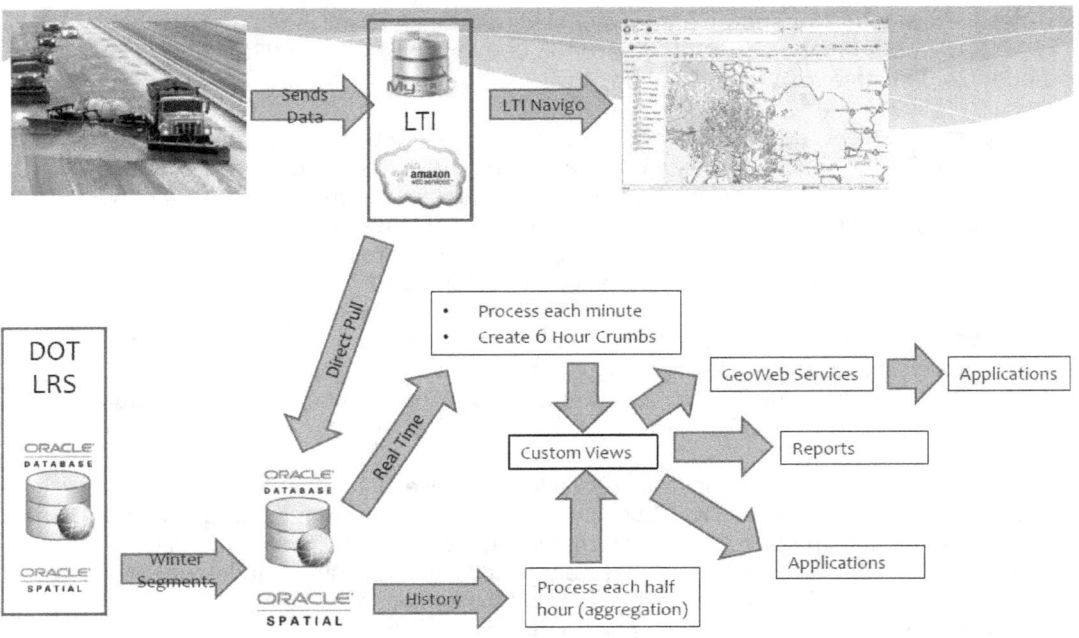

Figure 11: Data architecture
The data architecture of the Iowa DOT Snow Plow Portal starts with data collection (top-left) and ends with analysis (bottom-right).
Source: www.gis-t.org/files/vaRsw.pdf

While the underlying data collection and storage has demanded much of the agency's attention in the early stages of the tool's implementation, the web interface of the system has proven to be quite easy to build and manipulate. Iowa DOT's concerted effort to build a dynamic Oracle Spatial database for widespread agency usage has led to a strong familiarity among staff for coding and referencing data for use in web applications. Furthermore, the agency's focus on building a separate geospatial portal for its Highway Division has allowed it to experiment with the most efficient and cost-effective methods of building a GIS-based portal. These efforts led to Iowa DOT choosing Geocortex Essentials from Latitude Geographics as the front end for its GIS-based applications.

As the Snow Plow Portal is not yet completed, DOT staff members are still experimenting with the best ways to present information to users. Early efforts to build a highly customizable system for use by a variety of stakeholders have shown promise. Due to the power of the Oracle Spatial database, reports on any data characteristics that the agency collects can be pulled up and presented in the web-based service in short time. These data can be used to assist in real-time decision making, or even to ensure that a specific truck's components have been installed correctly. JavaScript maps have enabled supervisors to search for data on their mobile phones. Creating a real-time, data-rich, on-demand web application has proven to be successful for Iowa DOT even before the complete roll-out of the Snow Plow Portal.

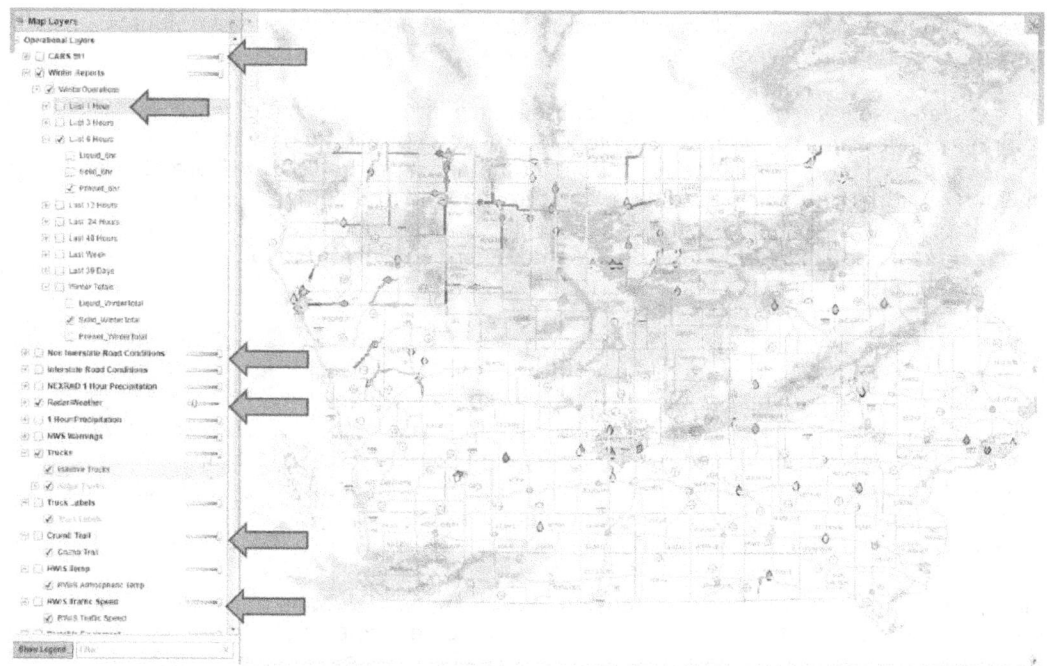

Figure 12: Snow plow portal with weather radar
Radar images are an available overlay for the Portal, assisting decision-makers in deciding when and where vehicles should be deployed over the course of a weather event.
Source: www.gis-t.org/files/vaRsw.pdf

From a long-term perspective, Iowa DOT can track the intensity at which materials have been used over the course of a weather event or an entire winter season. If supervisors spot a vehicle dumping more material than advised, it can notify the driver of this. Over time, the reduction of these instances may help bring about significant cost savings. The same information that can be evaluated for on-the-spot fleet deployment decisions can also be collected on a long-term basis to identify locations to target maintenance activities in future seasons. Particularly troublesome roadways may be identified for improvement projects.

Outreach

Outreach for this effort consisted of two parts. First, Iowa DOT staff members were tasked with creating a compelling argument for why management should invest in building the system. Given the technological complexity involved in assembling the digital and physical components of the system, the agency anticipated that it would encounter obstacles as it developed a complete system ready for use.

A Return on Investment study identified a $6.40 return for every $1 invested in the system. Given the budget restrictions facing Iowa DOT and the potential benefits of cutting winter maintenance activity costs, agency officials have deemed the Snow Plow Portal a high-level initiative within the agency. The director of Iowa DOT's Highway Division has been a strong proponent of the system, and part of the surplus budget from a mild 2012 winter season has been re-directed toward the Snow Plow Portal.

The second outreach factor involves on-the-ground discussions and training with Iowa DOT staff on how to best utilize the Snow Plow Portal. One of the most difficult arguments the agency has encountered is convincing mechanics and drivers that the system is not intended to over-scrutinize their job performance. The data and the metrics collected through the system aim to assist supervisors with management of fleet activities and promote an efficient use of resources, not to admonish mechanics and drivers for erroneous actions.

Acceptance of the system is improving and is anticipated to rise further as training activities become more prevalent. Iowa DOT expects to achieve widespread buy-in among its staff as technical staff members become knowledgeable about the database, more field staff are able to troubleshoot issues related to in-vehicle hardware, and the accurate and comprehensive collection of data is promoted. As the robustness of collected data cannot be ensured without properly trained technical and field staff, these developments are crucial for the long-term success of the system.

Benefits

Many of the anticipated benefits of the Snow Plow Portal will be derived from the power and flexibility provided by the ample data and customizable web applications. Supervisors and other stakeholders are already able to reference any number of data points to evaluate past performance, make future deployment decisions, and manage material usage. Iowa DOT expects the tool's influence to grow as more data is collected in future seasons and data collection becomes more standardized between vehicles.

These capabilities are made possible by the extensive Oracle Spatial database at the foundation of the tool. Iowa DOT familiarity with the database from use on other projects allows technical staff to easily reference data that supervisors request. The flexibility of both the Geocortex Essentials system and other web applications enables users to quickly and conveniently navigate the portal from their office or in the field. The integration of these web-based services with the foundation that is the agency's databases allows for a low-maintenance process on the front end of the portal.

Although the portal, which is still largely under development, has not yet led to documented achievements, supervisors have reported that they have been able to detect inefficient uses of resources during weather events. In one instance, a supervisor was able to notify the driver that material was not being spread properly and rectify the problem. As hardware becomes more standardized across vehicles and more winter weather events are encountered Iowa DOT expects to see the system provide many dividends for resource management.

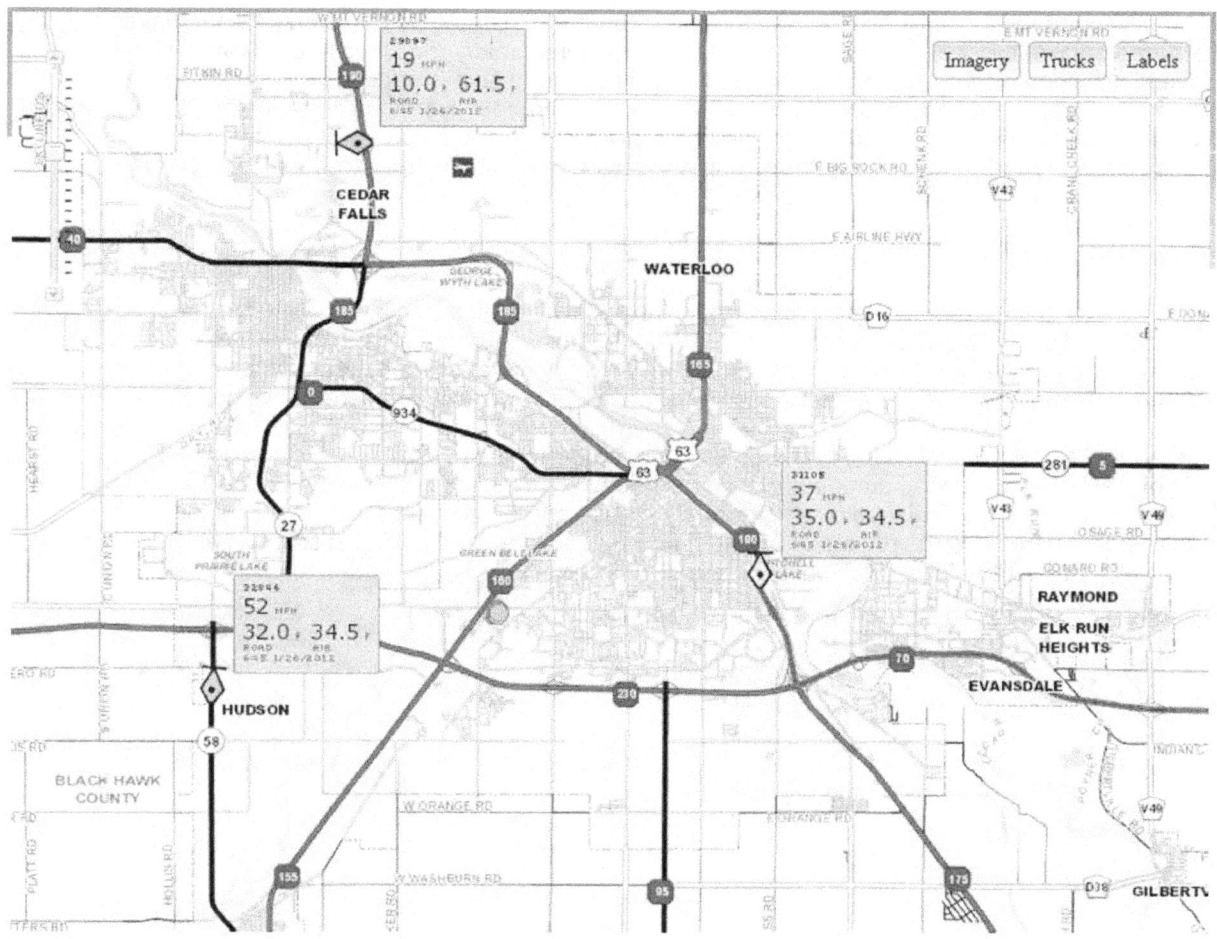

Figure 13: Real-time data
Staff can reference real-time data collected from deployed vehicles during a weather event, such as in this instance around the city of Waterloo. Data includes time, vehicle speed, air temperature, and road temperature.
Source: www.gis-t.org/files/vaRsw.pdf

Challenges and Lesson Learned

Building an innovative and technology-driven system that relies on sensor equipment in harsh outdoor conditions, has not come without its challenges. When the system was initially deployed in January 2011, each winter maintenance vehicle was equipped with a Windows computer in addition to LTI's custom radio modem. Each driver was expected to log into the system and report the conditions and events that her or her vehicle was collecting. This proved problematic due to numerous computer malfunctions, particularly with the cell air card plugged into each machine. Some drivers also found it difficult to input data while wearing gloves.

Of considerable concern was the inability of the sensor equipment to provide consistent and complete data throughout Iowa DOT's fleet. Components were found to have not been standardized properly across vehicles and temperature sensors were often inaccurate due to the heat from vehicle exhaust. Even detecting a truck's plow status has sometimes proven difficult, as the agency observed a number of instances for which a plow appeared to be on the road but was in fact raised.

For the winter of 2012, Iowa DOT began to remove on-board computers from vehicles and have real-time data sent to LTI's internal database. This has taken many responsibilities away from drivers, which many welcomed. The system has shown improved performance as a result of these changes, although a mild winter season prevented the agency from fully evaluating the system. Iowa DOT hopes to address continuing concerns in database performance and vehicle equipment standardization moving forward, but the agency is optimistic that the tool is close to realizing widespread benefits for winter operations.

Performance Metrics

Performance metrics for the Snow Plow Portal directly involve the potential cost savings the system may help attain. The average Iowa DOT budget for winter operations was $40 million between 2008 and 2010. Over this time, the agency did not have a strong understanding of how efficiently its winter weather materials, such as salt, calcium chloride, and sand, were being used.

The Snow Plow Portal tracks the real-time use of materials and can detect when too much or too little of a substance is being dumped. Additionally, road condition data is collected to evaluate how resources can be deployed more effectively. These efforts can lead to cost savings for the agency; a 10 percent in the reduction of salt used would lead to $1.4 million in savings, and potentially more in the future with price increases of materials.

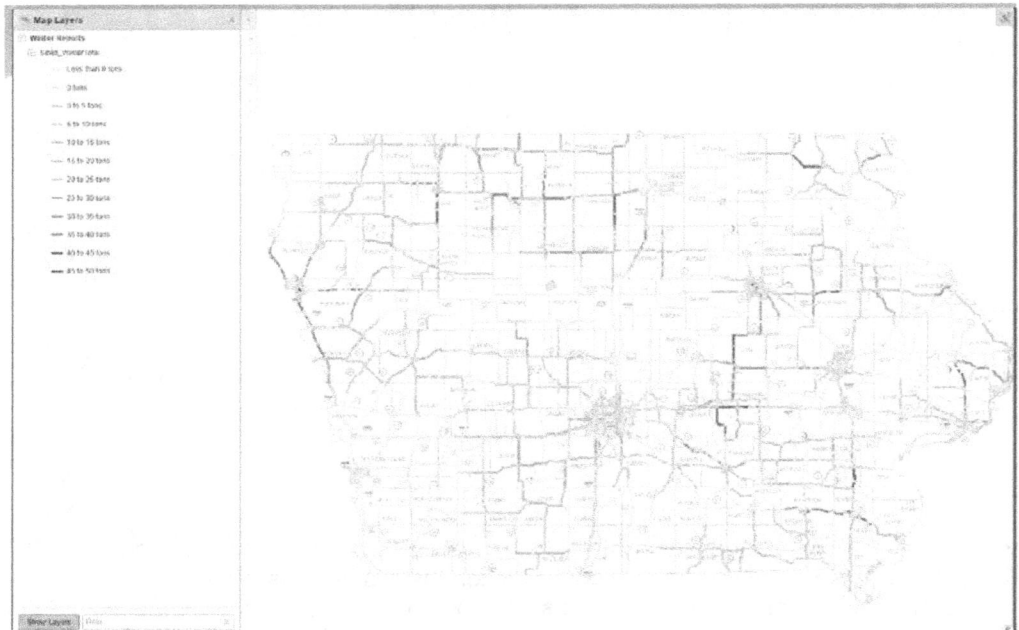

Figure 14: Analysis for the entire season
A screenshot showing the use of solid materials over the course of a winter season on Iowa's roadways.
Darker colors signify a higher usage of materials.
Source: www.gis-t.org/files/vaRsw.pdf

Future Direction

Iowa DOT has targeted a number of goals for the 2012/2013 winter weather season. From a data collection standpoint, the agency hopes to have most, if not all, of its plows equipped with sensor equipment. Additionally, it hopes to continue to improve the calibration of spreaders, temperature sensors, and other sensors. These improvements will help make data more comparable and comprehensive for analysis.

With data collection processes better executed, the agency believes that the system could be fully operational before the end of the next winter season. This will involve the introduction of a custom reports and custom portals for the different stakeholders interested in the data. As has been discussed, the strong database architecture for warehousing and processing within Iowa DOT allows the agency to easily link information for web-based analysis and generate data for reports.

The agency also hopes to eventually link the tool with its larger resource management system. The integration of these two systems will allow for analysis of data among broader audiences, as well as the improved assessment of winter operations cost factors in the context of total resource expenditures.

2.5 Kansas DOT's KGATE

Background

KGATE is web-based state transportation GIS portal developed by the Kansas Department of Transportation (KDOT). A product of collaboration between the department's Bureau of Transportation Planning and its Bureau of Computer Services, KGATE is accessible to all employees within KDOT. The tool, in conjunction with an enterprise data warehouse, has become an essential component of everyday operations within KDOT and currently plays a role in a range of decision-making processes throughout the agency.

Application

KGATE is comprised of 70 feature datasets from 13 unique sources, all of which are able to be viewed on or in conjunction with an online basemap. Much of the spatial data in KGATE come from internal sources, such as the state's data clearinghouse, the KDOT document management system, the KDOT pavement history research lab, and the Secretary of State's Office for Legislature Information. Other data come from publicly available sources, such as the FHWA Kansas Division Office, the Kansas Department of Revenue's truck routing system, and the Kansas Highway Patrol. KGATE combines all of this data into one tool that includes standard GIS point/line/polygon vector data, satellite and aerial imagery, videologs, documents, and photographs.

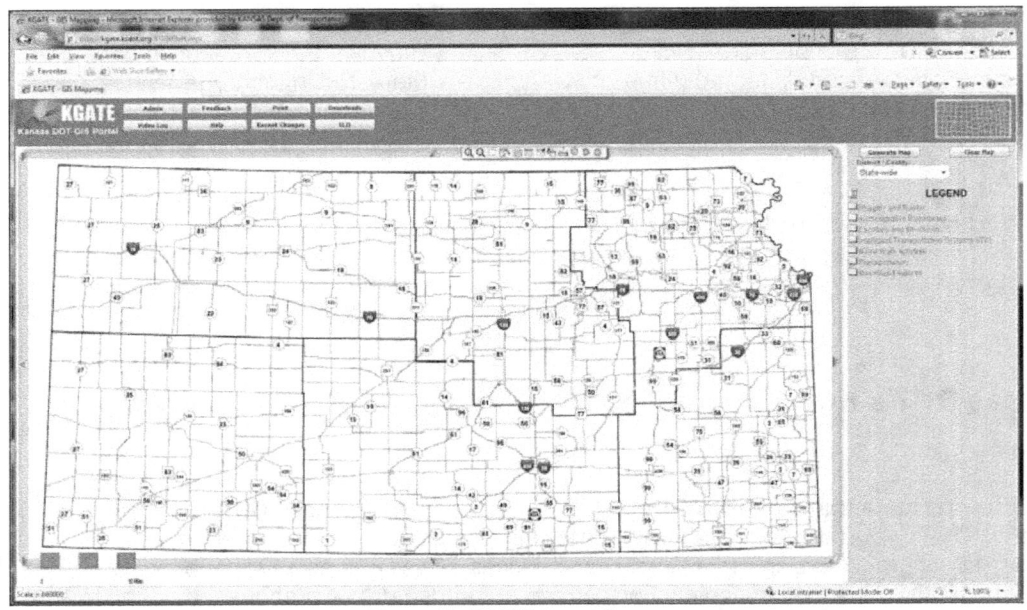

Figure 15: KGATE Home Screen
At full zoom, KGATE shows all major highway corridors in the state and the six state highway divisions. Application buttons are located at the top of the screen, with the legend and layer options situated on the right.
Source: www.gis-t.org/files/fjCZ2.pdf

Primarily a viewing tool, KGATE is not intended to support data manipulation. It is a single mapping application that allows users to compare many unique forms of KDOT data simultaneously, develop custom queries, generate reports tailored to their needs, and link to other applications (such as KDOT's videolog) that further illustrate the physical environment. An important KGATE user group is the KDOT Right-of-Way Bureau, which often uses KGATE as a reconnaissance tool to pre-assess conditions before visiting a location in the field. It is often used in conjunction with commercial 360-degree street imagery produced by Google Streetview or Bing Maps. The KDOT Traffic Engineering Group is another important user group that relies on KGATE when gathering data for federally mandated highway safety audits.

While the vast majority of KGATE's users are internal, the tool is occasionally opened to external users on a case-by-case basis after receiving clearance from KDOT Chief Counsel. An example is a consulting firm hired to redesign a highway corridor in a growing area of Kansas City. In this case, the consultant saves a significant amount of time by having immediate access to planning documents, crash data, and roadway geometrics. Other state agencies have sought access to KGATE, but the current leadership at KDOT prefers to keep closer tabs on transportation data usage outside the agency. Kansas has an open records statute in place which allows data requests to be monitored more easily.

KGATE uses Intergraph's GeoMedia Transportation and GeoMedia Web Map Pro for standard web mapping and network geometry.

Data Stewardship

KGATE's data is managed in an Oracle enterprise data warehouse, which based on its mission statement is "a flexible single source of integrated, reliable, and secure data that provides intuitive access to enterprise information supporting the Department's current and future business activities." The underlying motivation for the data warehouse is to emphasize data integrity and support a single version of the truth for all transportation data in the State. This allows KDOT to establish standards for data quality and conformity while facilitating data integration at high performance levels. To further minimize unauthorized use of data, KDOT does not make its data available for download via KGATE, and there is no export capability beyond copying tabular data and pasting it into a working spreadsheet.

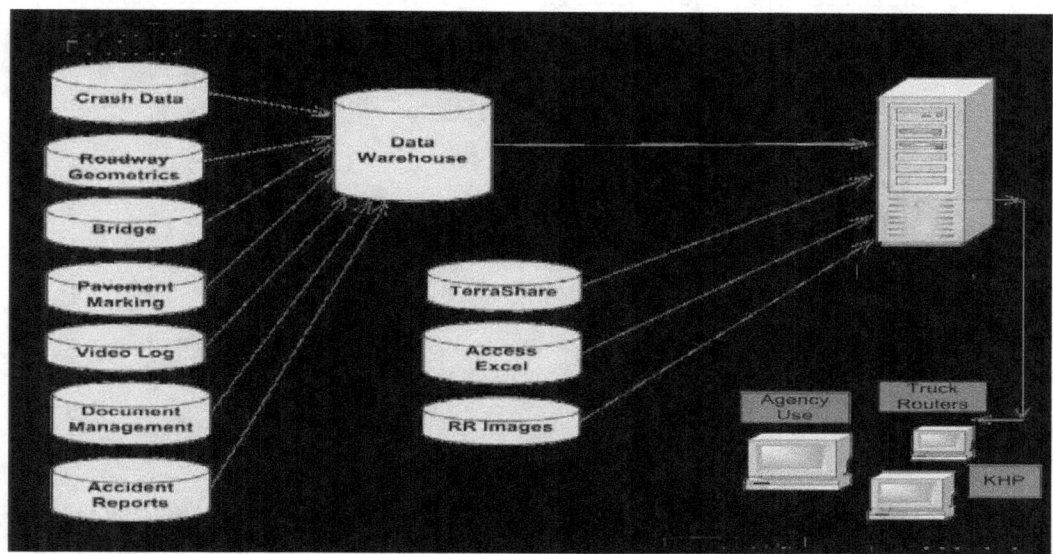

Figure 16: KGATE Data Flow Diagram

Much of KGATE's data is housed in the data warehouse, but a few items have yet to be incorporated or come from other sources outside KDOT.

Source: www.gis-t.org/files/fjCZ2.pdf

KGATE is a tool that benefits from data that are sourced from a variety of different production databases. Much of this data is housed in the enterprise data warehouse (crash data, roadway geometrics, bridge data, pavement marking, video documents, and accident reports), but KGATE also pulls information from internal and external sources (Figure 2). Most importantly, the KGATE team is not responsible for creating or maintaining any of these data; instead, those who are most familiar with the data are responsible for keeping it up-to-date.

Outreach

The KGATE team does not simply build tools and expect people to use them; rather, tools are built when a need has been identified and a business case for the effort has been presented to and funded by KDOT leadership. In order to determine the functional needs and direction of KGATE, the team traveled throughout KDOT to determine needs that are currently being met and opportunities for improvement or new functionalities. To date, the team has gathered input from many different user groups, including planners, traffic engineers, surveyors, intelligent transportation systems (ITS) specialists, and public information providers.

KGATE is intentionally a straightforward, user-friendly tool. While some users are aware of the analytical capabilities of viewing multiple types of data at the same time, many KGATE users still only wish to work with one dataset at a time. The development team therefore seeks to build new functionality where appropriate, while ensuring that users are not confused by extraneous data that may clutter or inhibit the core functionality of the tool. KDOT believes that many of its web-based map users are task-driven and apprehensive about tackling new technologies. Furthermore, it is difficult to find time and funding for extensive training sessions.

Support

The leadership at KDOT is widely supportive of KGATE, from the State Secretary of Transportation to the Bureau directors to the department managers. KGATE been dubbed "the GIS tool for the ages" by KDOT leadership, and while the GIS team is judicious about its funding requests, it has generally had success securing funds for the tool's development. Much of KGATE's success is likely attributed to the fact that it is a collaborative effort focused on delivering a functional product that benefits KDOT workers across the agency.

Performance Metrics

Although KGATE does not have a formal system in place for measuring performance, the KGATE team has recently added Web Trends capability to get a clearer understanding of the number of KGATE users. Recent results show that at least 300 KDOT employees regularly use the tool.

KGATE does occasionally solicit user feedback through surveys or focus groups, with its most significant effort just underway as planning begins for a new version of the tool. See the Future Plans section below.

Lessons Learned

According to the project managers, the most important lesson learned from developing KGATE is to emphasize collaboration. Because it is a joint effort between two KDOT bureaus, KGATE's developers always worked with input from both the Computer Services side (IT specialists, data managers) and the Planning side (map users and analysts). By continually working together and bringing both groups together for meetings and training opportunities, the collaborative spirit permeated all components of the tool.

Future Plans

As KGATE evolves and KDOT develops plans for future improvement, the KGATE team is in the process of gathering feedback and soliciting suggestions for the newest version of the tool. Much of this effort revolves around seven focus groups comprised of users in various departments throughout KDOT. Participants were selected based on a survey that helped identify KGATE users and their map-related tasks. The process also involves a business requirements group, which is charged with compiling and refining feedback from the focus groups and developing recommendations for the most cost effective solutions. Finally, a steering committee at the Bureau Chief level will create a business case for making improvements to the tool and champion the investment throughout KDOT. When new plans are finalized, they will pass through a hierarchy of committees in the IT Department, who will ultimately approve the business case. While the process is quite lengthy, it ensures that a large portion of KDOT is aware of KGATE's future plans and understands the funding needed to execute the desired improvements. It is likely that this process will result in a three-year or five-year plan for KGATE.

The KGATE team also aims to continue refining the tool and maintaining business processes that result in clean and reliable information upon which accountability can be maximized. Areas of focus include consistent metadata development, improved collaboration with the KDOT legal team, optimized performance and data processing speed, and data accuracy. While continuing to grow and improve KGATE, KDOT is ensuring that GIS fundamentals are in place and a solid foundation is in place for all Kansas' spatial transportation data.

3. Conclusions and Lessons Learned

GIS portals built by transportation agencies can serve a wide variety of needs, both internal to an agency and external to the general public. Some transportation agencies use GIS portals as a way to manage and query data pertaining to an agency's transportation assets, or for a specific subset of assets (such as those by a type of mode). A centralized, data-rich GIS application can assist agency employees in querying relevant information related to user-defined characteristics of a transportation system, such as the year of construction for all railroad bridges in a county. This type of tool can help track the conditions of organizational assets, manage maintenance needs, prioritize investment decisions, and address questions from the public or lawmakers. In practice, GIS portals can be complex and resource-intensive to maintain, requiring powerful databases and knowledgeable staff for support. However, a high-quality system can greatly improve an agency's internal operations.

Asset management systems, such as those for bridges, highways, and pavement, can also be integrated with GIS portals to assist with analysis. These systems are built to help manage the conditions of a specific asset within the transportation system. They can make decision-makers aware of when a component of the system has fallen beneath an established measure of safety or acceptance. In this instance, a portal can help an agency visualize condition information to allow for easier analysis of potential remediation actions.

GIS portals have shown promise in assisting the general public with making transportation decisions. A common use of a GIS portal has been in disseminating traffic information to travelers. Real-time traffic, accident, and weather information feeding into a web application can help direct travelers away from areas of congestion or construction. This can be helpful during the traditional weekday rush hours when there is excess demand for roadways, as well as in the event of a construction delay, accident, or severe weather event, when travelers can be advised to avoid specific routes of travel. This type of application also extends to other modes of travel, such as notifying commuters of train and bus departure times.

A more data-intensive example of a GIS portal for public consumption is the use of interactive base maps, which allow users to query information relating to the transportation system. Transportation data may be combined with information pertaining to other topics to allow engaged citizens or business interests to carry out work. Examples of this include overlaying transportation networks over endangered natural habitats or pollution levels to study the influence of vehicle emissions on these subjects. To cite one example in business, retail analysts may refer to historical traffic counts to argue for the location of a retail outlet at a particular location.

3.1   Essential Portal Components

There are three essential components of a GIS portal: a database, a GIS server, and an Internet application framework.

Database

The database is the foundation of a GIS portal and is the key component for data storage and organization. The portals discussed in this report require powerful data management systems that store large quantities of data and quickly process complex user requests. Every GIS portal covered in this report uses Oracle Spatial, an add-on to the Oracle Database Enterprise Edition that supports high-end GIS and location-based services. Other data management tools include SQL, a programming language designed for relational databases, and Microsoft SharePoint, which is helpful for managing documents requiring varying levels of security.

GIS Server

Standard GIS applications serve as a way to store, analyze, and present geographic data. Many of the capabilities of GIS portals are made possible through functionalities of GIS applications, and many transportation agencies have chosen enterprise GIS programs that rely on long-term agreements with software providers.

ESRI's ArcGIS Server and Intergraph's GeoMedia WebMap Professional are two of the major players in the development of GIS servers that support interoperability across multiple systems and platforms, including applications for desktop computer, mobile devices, and the Internet. Implemented on-site or in the cloud, a GIS server enables agencies to create highly customizable web applications for both internal and external use. An API (application programming interface) serves as the translator between traditional web-based GIS functionality and the web-based functions that are often incorporated into a GIS portal.

Internet Applications

A web application processes and manipulates data fed into it by GIS to allow for outward-facing consumption by external users. In this instance, web applications enable users to move across different geographic areas, specify data to display, and query data made available by the provider. In a broader context, web applications allow for the integration of many dynamic sources of multimedia while using the open web browser as a client, such as downloading documents, submitting forms, playing video, and even making online purchases.

3.2 Lessons Learned

Upon conducting interviews and drafting case studies, four important lessons emerged. Each is expected to play an important role as agencies continue to develop GIS portals in the coming years.

Focus on the User

Above all, every GIS portal profiled in this report seeks to be user-friendly and task driven. From the specific Oversize/Overweight Vehicle Pre-Screening Tool of New York State DOT to the more comprehensive portals developed by the Kansas DOT (KDOT) and Georgia DOT (GDOT), ease-of-use is paramount. Each agency makes a strong effort to provide outreach support, conduct training, design intuitive interfaces, solicit feedback, and empower users with functionality that meet their stated needs.

Each GIS portal examined as part of this report was created with a targeted set of desired outcomes. Portals exist to simplify a business process, achieve results, reduce extraneous data, and minimized or eliminate the learning curve for users without prior GIS experience. The KYTC's interactive mapping program offers anexample of this theme, developing wholly distinct portals tailored closely to the needs of specific user groups. With its use of focus groups, KDOT's KGATE portal ensures that users' needs are met and their desired functionality is achieved.

Institutional Support

Over time, GIS has become a needed tool for transportation organizations to carry out their mission. The concept of a GIS portal, intended for more widespread use by employees and the general public, has mirrored this growth. The portals profiled in this report enjoy firm institutional support from the administrations of their agencies. These parties recognize the value and importance of developing these tools to improve productivity, support planning efforts, and communicate with the general public. Thus, labor and other resources needed to build these systems are pledged from the beginning, and setbacks are tolerated.

Data Stewardship

As GIS data becomes more prolific, transportation agencies are designing their GIS data architectures in a way that specifies interdepartmental roles with regard to data collection, management, and dissemination. For teams working on GIS portals, emphasis is placed on the importance of not creating data. That is, a GIS portal collects, processes, and disseminates data. Those within the agency who are most familiar with the data are thus responsible for its creation and maintenance. For example, pavement specialists are responsible for collecting information about pavement conditions; wildlife specialists are responsible for incorporating changes to habitat maps; bridge engineers are responsible for reporting changes in bridge condition status. According to the case study examples, ideally, every data layer has a steward who keeps the information up to date and is able to address questions and/or issues with the data raised by tool users.

As a result of data stewardship programs, the team responsible for developing and maintaining a GIS portal is able to concentrate on the flow of information from the field to the database to the desktop computer or handheld device.

Performance Metrics

None of the DOTs interviewed for this series of case studies have robust systems in place for measuring the performance of its GIS portal. While most agencies do track web statistics and distribute surveys to solicit feedback from users on the successes and needed improvements of the tool, there are little quantitative data to suggest that GIS portals are generating a return on the investment. Potential opportunities for quantitative measures include cost savings incurred from the release of a certain portal functionality or feature, calculated time savings related to data collection processes for federal documentation requirements, and fuel savings resulting from a reduced number of trips into the field.

Appendix A: Interview Guide

Background / Purpose
1. Please provide a short overview of your tool. When and why was it developed? What is its current development status?
2. Do you consider it a dashboard? A portal? Something else? Do have an opinion as to the distinction between these types of tools?
3. What is its primary purpose? Secondary purposes?
4. Who is the primary audience? Do you know of other user groups that benefit, intentionally or otherwise?

Software and Data
1. What software platform(s) are you using? What were your options? What were your reasons for choosing one product/technique over another?
2. What types of GIS data are incorporated into your tool? Do you have a process for deciding what data to include/exclude?
3. Does your agency collect this data for purposes other than your tool? Does it have to be specially formatted for this application?
4. What are your data sources?
5. From an IT perspective, how is your data/tool integrated into other programs (GIS or otherwise) within your agency?
6. From a non-technological perspective, how is your data/tool integrated into other programs within your agency?
7. In terms of labor and funding, what resources are required to operate the tool and/or collect data? Do you have a budget in place for this tool?
8. Do you have a plan in place for data maintenance/updates?

Application and Decisionmaking
1. How is your GIS dashboard or portal used? How is it accessed? Do you know if it is actively being used? By whom?
2. Is the data for viewing, downloading, and/or manipulating? Why?
3. Do you know of any specific effects that your tool has had on planning, operations, and or maintenance within your agency? At the working level? Management level?
4. What are your future plans for the tool? How do you make decisions about what to tackle next?
5. Has the tool led to any documented benefits by a user from the general public? Has the tool been cited as helping to bring about a specific achievement?

Evaluation
1. What challenges/obstacles have you or your agency encountered? How have they been addressed?

2. How have you funded the development of this tool? Did you encounter and/or do you foresee any major funding challenges?
3. Do you offer programs, such as outreach, marketing, or training opportunities, that support or utilize your tool?
4. Is there a strategy for evaluating and/or measuring effectiveness of the tool?
5. What has been the response to the tool – either internally, from the public, or others? Have your efforts been documented?
6. Can you provide examples of reports you produce or have produced for internal or external communication purposes?
7. What would you say to another transportation agency considering a similar venture?

Other/General
1. Any more information you would like to add?
2. Any other documents/literature we should review for more information about the tool?
3. Can you recommend any other contacts?
4. What one thing would you do with your GIS dashboard or portal program if there were no constraints?